# SHARKS

## Threatened Ocean Predators

Hal Marcovitz

ReferencePoint Press

San Diego, CA

ReferencePoint Press®


Printed in the United States

**For more information, contact:**
ReferencePoint Press, Inc.
PO Box 27779
San Diego, CA 92198
www.ReferencePointPress.com



LIBRARY OF CONGRESS CATALOGING-IN-PUBLICATION DATA

Author: Hal Marcovitz
Title: Sharks: Threatened Ocean Predators
Description: San Diego, CA : ReferencePoint Press, 2026.
Includes bibliographical references and index
Identifiers: LCCN 2025006209 (print) | ISBN
9781678211967 library binding | ISBN 9781678211974 ebook

For compete cataloging-in-publication data please go to www.loc.gov.

# CONTENTS

INTRODUCTION

# The Challenge of Saving Sharks

Tamayo Perry was a local celebrity well known across the Hawaiian island of Oahu, where he made his home. He worked as a lifeguard on the beaches of Oahu. He also taught surfing—an enormously popular sport in Hawaii as well as many mainland beach communities. Perry was an expert surfer, often participating in competitions in Hawaii and other surfing hotspots in the world. And he occasionally found work as actor, cast in small roles in movies and TV shows that are filmed in Hawaii. "Besides surfing, another huge passion of mine is acting," Perry wrote on his personal website. "When I'm not acting, I'm still quite a character, so there's always a ton of fun to be had."[1]

On June 23, 2024, Perry was surfing near Goat Island, a tiny islet just off the northeastern edge of Oahu, when he was attacked by a shark. The attack proved fatal to the forty-nine-year-old surfer. When lifeguards reached Perry's body, they discovered he had suffered numerous shark bites. Jesse King, a close friend and fellow lifeguard, said Perry knew the Goat Island waters very well and would have had no reason to believe his life would be threatened by a shark attack. "There was nothing out of the ordinary that would indicate that something bad would happen to him. . . . A regular day,"[2] says King.

Shark attacks are not common in the waters surrounding Hawaii, but they do occur occasionally. In the twenty years prior to Perry's death, only eight other swimmers and surfers had lost their lives to shark attacks not far from the beaches of Oahu and the other Hawaiian islands.

## Shark Attacks Are Rare

Fatal shark attacks are rare around the globe. According to the Florida Museum of Natural History, which tracks shark attacks around the world, in 2023 there were sixty-nine "unprovoked" shark attacks on surfers, swimmers, fishers, and others across the globe. This statistic reflects the typical number of shark attacks that have occurred annually in previous years. An unprovoked attack typically occurs when swimmers, surfers, scuba divers, fishers, or others suddenly and unexpectedly find themselves attacked by sharks. Among the sixty-nine unprovoked attacks in 2023, ten were fatal.

In addition, the museum reported that in 2023 there were twenty-two "provoked" attacks on humans around the world. This means the victims approached the sharks in some fashion—perhaps in a friendly attempt to offer food to the sharks or to reel in sharks caught by fishing lines. In the provoked attacks, four victims were killed by the sharks.

To swimmers, surfers, and others, sharks may appear to be fearsome predators, but in reality, sharks are more often the victims of human activities than the other way around. According to a 2024 study authored by several marine biologists and published in the journal *Science*, some 80 million sharks a year swimming across the world's oceans are killed by humans—and sport and commercial fishing are usually to blame. In fact, the annual catch of sharks by fishers is so huge that many types of sharks are now regarded as either threatened or endangered species—designations that call attention to the likelihood that those species may soon be lost to extinction. "If a shark is ever to encounter a person it is far more likely that the person will kill the shark than the shark will kill

*Surfer Tamayo Perry, pictured here at the 2005 WCT Billabong Pro competition in Teahupoo, Tahiti, lost his life in a 2024 shark attack.*

the person," says marine biologist Jasmin Graham. "They have every reason to be afraid of us; we have very little reason to be afraid of them."[3]

## Mission to Save the Sharks

In America and other countries, marine biologists, environmentalists, and many government leaders hope to find ways to save the sharks—to reverse the trend that has threatened to wipe out numerous species of sharks. In the meantime, though, shark attacks on people may be rare, but they continue to occur. Each incident adds to the popular belief that sharks are fearsome and relentless predators and that people who hope to enjoy their days at the beach would do well to stay out of the ocean. In 2023 ten-year-old Dillon Armijo and his family were vacationing in Cancún, a seaside resort city in Mexico. Dillon and his

> **"If a shark is ever to encounter a person it is far more likely that the person will kill the shark than the shark will kill the person. They have every reason to be afraid of us; we have very little reason to be afraid of them."[3]**
>
> —Jasmin Graham, marine biologist

brother were wading in the knee-deep ocean waters just a few feet from the shore when, suddenly, Dillon was attacked by a shark. "It just felt like a bump, like a really sharp bump . . . just like a big wave that hit me," Dillon said later. "And I just didn't see anything. I just saw the blood."[4] Dillon's brother and a lifeguard carried Dillon onto the beach, where the lifeguard applied a tourniquet. Dillon was taken quickly to a nearby hospital, where he was treated for deep wounds to his knee. Later, Dillon returned home to Colorado, where he has undergone several surgeries to repair the damage to his knee. Doctors expect Dillon to eventually make a full recovery.

> **"I just didn't see anything. I just saw the blood."[4]**
>
> —Dillon Armijo, shark attack victim

The attacks on Perry, Dillon, and others have helped cement the notion that swimming just offshore or riding the waves atop a surfboard may be dangerous activities. "If you're going to get in the water, you have to assume the risk," says marine biologist Neil Hammerschlag. "But I think people should take a kind of comfort in knowing that a shark bite is extremely rare."[5] Nevertheless, it is hard to imagine that the attacks that killed Perry and severely injured Dillon would engender much sympathy for the sharks—meaning the mission to save the global shark population will undoubtedly find itself following a challenging path ahead.

CHAPTER ONE

# The Unique Qualities of Sharks

Sharks are easily recognizable. They are typically very large fish—most adult sharks measure 3 feet (91 cm) or longer in length. Sharks are sleek, with bodies colored blue, brown, olive, black, or gray. Most sharks feature jawlines that turn downward—giving them a decidedly angry and threatening look. They can be very fast swimmers. According to the Washington, DC–based Smithsonian National Museum of Natural History, the great white shark can swim as fast as 35 miles per hour (56 kph). In contrast, Michael Phelps, who won twenty-three gold medals in his career as an Olympic swimmer, averaged about 5 miles per hour (8 kph) during his competitions.

Sharks are well-known creatures of the seas because they have been around for a long time—much longer than humans. The ancestors of today's sharks date back some 450 million years. The first sharks populated the oceans even before the first dinosaurs thundered across the world's plains and forests. In fact, the dinosaurs did not show up until some 255 million years ago. Moreover, the dinosaurs were extinct by about 65 million years ago—but the sharks survived and continued to thrive in their undersea world. In contrast, it has only been for about the past two hundred thousand years that the modern human species has existed on the planet.

Therefore, there is no period in human history in which sharks have not been lurking about offshore, no doubt sending chills of fear through the first humans who waded into the sea or piloted the earliest and crudest boats across the waves. Says marine biologist Greg Skomal:

> Sharks have prowled the world's waters for the last 450 million years. Compare that to modern humans who have only been around for about 200,000 years! It is remarkable to think that some of the shark species . . . that swim in our oceans today were around when dinosaurs roamed the earth. What is even more amazing is that the dinosaurs became extinct about 65 million years ago, but the sharks persisted and still dominate the watery world.[6]

## Bundles of Cartilage

There are more than five hundred known species of sharks. Several characteristics unify all species of sharks. The one characteristic that links all sharks is the fact that they have no bony skeletons. Instead, their anatomies are surrounded by bundles of cartilage—the same soft and pliable tissue that makes up people's ears or the tips of their noses. Another common characteristic found in all sharks is a very powerful tail fin—giving them the ability to be fast swimmers. This gives them an advantage when they target their prey. Moreover, the bodies of sharks are covered in sleek V-shaped scales known as denticles. The denticles reduce water resistance while the sharks are swimming—another speed enhancement that makes it easier for the sharks to glide through the water. (Water can pull and tug at bodies, slowing a swimmer's ability to move through it. Competitive swimmers, for example, typically wear rubber bathing caps because the hair on their heads causes resistance to water—in other words, their hair slows them down.)

**"Sharks have prowled the world's waters for the last 450 million years. Compare that to modern humans who have only been around for about 200,000 years!"[6]**

—Greg Skomal, marine biologist

Finally, all sharks have what are known as revolver jaws—meaning that throughout their lifetime, their teeth are constantly replaced as they wear out. Some species of sharks may have as many as fifty thousand teeth during their lifetimes. In contrast, humans start receiving their first set of teeth—baby teeth—when they reach age four to seven months. When humans reach age six to seven years, their permanent teeth begin replacing their baby teeth, and those teeth have to last for the rest of their life.

The hundreds of species of sharks may share many common characteristics, but there are also a lot of differences found from species to species. Size is one such characteristic. The tiniest shark is the dwarf lantern shark, which is smaller than a human hand; the largest is the whale shark, which is about 40 feet (12 m) in length, or about the size of a school bus. There are many other physical features that differ from shark to shark. Hammerhead sharks, for example, have flattened heads and eyes that

*Great white sharks are among the world's most feared creatures. These predators can pursue prey at speeds of up to 35 miles per hour.*

protrude from the sides of their snouts. Leopard sharks have oval spots across their bodies—much like the leopards found in the African jungles. The bigeye thresher shark is so named because of its large eyes—each eye is typically about 4 inches (10 cm) in diameter—and it's long, scythe-like tail fin.

## Abandoning Their Pups

Whether they have big eyes or leopard-like spots, there are characteristics linking sharks that go beyond their denticles, bundles of cartilage, and powerful tail fins. For example, unlike other fish, many shark species are not restricted to either salt water or fresh water—meaning sharks can be found in saltwater oceans as well as freshwater rivers. The bull shark, for example, is usually found swimming in the world's saltwater oceans near tropical shorelines but has also been seen swimming in freshwater rivers, including the Mississippi River in the United States and the Amazon River in South America.

Many shark species have a lifespan of hundreds of years. The Greenland shark—so named because it can be found swimming near the Arctic island of Greenland—has a lifespan of some four hundred years. This means there are likely Greenland sharks still living today that were swimming in the sea when the American colonies rebelled against the British monarchy in 1776.

Shark pregnancies typically last about a year, but some female sharks—among them the frilled shark and the basking shark—carry their fetuses for as long as three years. Some sharks, such as mako sharks and bull sharks, give birth to live offspring, but other sharks, such as cat sharks and bullhead sharks, lay eggs. Regardless of how they birth their young, sharks are not mammals. All sharks are fish.

Sharks that give birth to live offspring immediately abandon the pups (as they are known), leaving them on their own to find food and fend off predators. But for the baby sharks, it is not as bad as it sounds. Female sharks nearing the birth of their litter are known to swim toward shorelines, where they give birth in shallow

## Some Sharks Do Not Swim While They Sleep

Although most sharks remain in motion while they sleep, some species remain still while they sleep, typically coming to rest on the ocean floor. These sharks are known as buccal pumping sharks. (*Buccal* refers to the shark's cheeks.) These sharks open and close their mouth while sleeping. This action forces water into their gills through their inner anatomy. As the water passes through their gills, their bodies extract oxygen from the water, which keeps them alive.

A species of shark known as the nurse shark, found near Florida and in other warm water regions, is a buccal pumping shark. Reports the British-based environment group Shark Trust:

> In essence, this method uses the cheek muscles to draw water into the mouth and force it out again over the gills. This keeps the gills oxygenated even when the shark isn't swimming. Some species, like hammerheads, have to keep swimming continuously in order to supply oxygen to their gills, otherwise they would suffocate. The buccal pumping method allows nurse sharks to live a more sedentary life resting on the sea floor during the day and being more active at night.

Shark Trust, "Creature Feature: Nurse Sharks," 2020. www.sharktrust.org.

waters as a degree of protection for the pups. Predatory fish—including many adult sharks—are less likely to go hunting for food near shorelines. Such areas are known as shark nurseries. It is possible to find the offspring of several mothers populating the shark nurseries. After their birth, the young sharks remain in the nurseries for several years. As they mature into adult sharks, they leave the nurseries and head for deeper waters.

### Risks to Hatchlings

As for sharks that lay eggs, for the mother shark the experience bears little resemblance to how birds lay eggs, which are then protected and nurtured by the mother bird as they hatch and tiny chicks are born. When the typical bird prepares to lay her eggs, the mothers—and often with the assistance of the fathers—fashion nests in trees. Nests provide a place for the mothers to lay their eggs and keep them safe and warm until they hatch. The mothers and frequently the fathers of the chicks then bring food back to the nest, nurturing the chicks until they fly off on their own. But when mother sharks are ready to lay their eggs, they deposit them on the seafloor, where they hatch on their own. The baby sharks emerge from their eggs and then grow and mature with no help from their mother. But again, for the embryos growing inside those shark eggs, it is not as bad as it sounds.

When, for example, a bullhead shark lays her eggs, she immediately pivots around and takes the eggs into her mouth. She then carries them to the ocean floor to search for crevices or cracks in the rocks where they can remain hidden from predators. And then she departs, leaving the eggs to hatch without her help.

Despite these efforts, some eggs will become food for predatory fish. That is why the typical shark lays as many as two hundred eggs. With litters this large, it is likely that many of those eggs will hatch. And while many of the hatchlings will

*The shortfin mako has sharp, backward-pointing teeth that are ideal for grasping fish and squid. A single shark may lose and regrow up to fifty thousand teeth over its lifetime.*

undoubtedly become victims of predatory fish, many will survive and grow into adult sharks.

For some shark embryos, their paths to birth are quite different. Some species of sharks lay eggs, but the eggs remain within the mother's body until they hatch. These litters are typically very small, usually containing no more than eight eggs. When the first eggs hatch, the pups, still in their mothers' bodies, are likely then to devour the other eggs. Says Melissa Cristina Márquez, an Australia-based marine biologist, "It's a shark-eat-shark world: literally! It never ceases to amaze people that one of the ocean's top predators is fine taking a chunk out of their own species. . . . But did you know that shark-on-shark eating happens inside the mother shark as well?"[7] Among the sharks that carry eggs within their body, giving birth to pups that devour the other eggs, are the great white shark and tiger shark—two species that are among the most fiercely predatory fish swimming beneath the seas. And given the fact that these sharks begin their new lives by devouring their own siblings, it is easy to see where their predatory habits begin.

## A Shark's Diet

However the pups may begin their lives, once they leave their mother's body or their broken egg shells on the ocean floor, they immediately begin to search for food. When it comes to finding food, sharks have some unique habits. All sharks feed on other living creatures below the ocean surface. A typical example is the bonnethead shark, which preys on lobsters, shrimp, and crabs. But bonnethead sharks, as well as many other species of sharks, also feed on seagrass that grows along the ocean floor. Therefore, the bonnethead and many other sharks are omnivores—they eat both meat and plants.

Another omnivorous shark is the whale shark. The huge shark weighs as much as 20 tons (18 metric tons). Despite their fearsome size, whale sharks are not regarded as predators—

*Whale sharks, which can reach about 40 feet in length, are the largest shark species. These gentle giants eat only small creatures and are harmless to humans.*

that is, attacking other fish. Rather, whale sharks feed by opening their huge mouth, then sucking into their bellies whatever they may find in their path.

Typically, their diet includes fish eggs, small fish, small squids, tiny shrimp-like creatures known as prawns and krill, and near microscopic sea creatures known as plankton. Whale sharks also eat seagrass. A 2019 study by marine biologists at the University of Tokyo in Japan included taking blood samples from whale sharks. Analysis of those samples revealed that half the diet of a whale shark is composed of plant life. Says marine biologist Jasmin Graham, "We have giant sharks like the whale sharks that are just massive and huge and what do they eat? Tiny little plankton. And they have throats the size of golf balls. They could not swallow you if they wanted to."[8]

**"We have giant sharks like the whale sharks that are just massive and huge and what do they eat? Tiny little plankton."[8]**

—Jasmin Graham, marine biologist

## History's Largest Shark

The largest shark to ever swim in the earth's oceans was the megalodon, which lived some 20 million to 3.6 million years ago. The megalodon—which resembled the modern great white shark—was believed to span some 60 feet (18 m) in length. The modern great white shark typically reaches about 15 feet (4.6 m) in length.

The shark's name means, literally, "giant tooth," and it is easy to see why. Scientists have recovered megalodon teeth that have measured about 7 inches (17.8 cm) in length. Moreover, the megalodons were apex predators of their era—devouring fish, whales, and anything else in their path. Marine biologist John Long describes what was likely the typical megalodon attack:

> The massive 60-foot shark crashed through the gloomy waters to charge upward and breach the water, seizing [a whale] in its powerful jaws. . . . When coupled with the incoming momentum of the shark's massive body weight, the jaws rammed the prey, as the beast crashed down into the water, with one powerful bite sliced the defenseless whale into two halves. Thick red billows of blood darkened the water. With a sharp turn of its head, the gargantuan shark swooped to finish the job, gulping the tail and body of the whale.

John Long, *The Secret History of Sharks: The Rise of the Ocean's Most Fearsome Predators.* New York: Ballantine, 2024, p. 309.

## Apex Predators

Swimmers and surfers may have nothing to fear from whale sharks, but there is no question that many other species of sharks are predators. In fact, the great white shark and tiger shark are known as apex predators, meaning that no other creatures in the sea hunt or prey on them. In other words, they are at the top—the apex—of the food chain. Since they have no natural predators, apex predators are free to roam wherever they like in search of food.

When sharks search for prey, they rely on the same senses that humans and many other creatures possess: sight, hearing, touch, smell, and taste. Sharks also rely on a sixth sense for this important task. Through a series of pores that are spread across the snout of a shark, it can detect very tiny electrical impulses. These impulses are known as nanoscopic electromagnetic currents. These impulses are emitted by living organisms, including fish and other sea life, as well as humans. In humans, the movement of limbs or other muscles generates an electrical field around the person's body; on land, the field immediately dispels because the atmosphere is not a good conductor of electricity. However, seawater is an excellent conductor of electrical current.

This sixth sense in sharks was discovered in 2018 by scientists at the University of California, San Francisco. Moreover, their study found that when a shark detects a nanoscopic electromagnetic current, its first impulse is to attack. The study found that the electrical impulses trigger the release of neurotransmitters—chemicals in the blood that spark reactions in the brain. These neurotransmitters tell the shark a meal is nearby. Said University of California, San Francisco, physiologist David Julius, who studies how living organisms function, "Sharks have this incredible ability to pick up nanoscopic currents while swimming through a blizzard of electric noise. Our results suggest that a shark's electro-sensing organ is tuned to react to any of these changes in a sudden, all-or-none manner, as if to say, 'attack now.'"[9]

> **"Our results suggest that a shark's electro-sensing organ is tuned to react to any of these changes in a sudden, all-or-none manner, as if to say, 'attack now.'"[9]**
>
> —David Julius, physiologist

Research has found that although all sharks possess the pores that detect electrical current, the apex predators have more pores than sharks known to be omnivores. Apex predators such as great white sharks and tiger sharks have as many as fifteen hundred pores across their snouts; other sharks have fewer—no more than a few hundred. Scientists suggest this might be a reason to fear apex predators: they have more pores, giving them more of an ability to find the sources of electrical current than other sharks that do not ordinarily target live prey only.

## Eyes Wide Open

But even apex predator sharks are not always on the prowl. It is not unusual for a swimmer to notice a shark passing by that clearly has no interest in attacking nearby prey. It may be that the shark is asleep.

For generations, sailors insisted that sharks do not sleep—that whenever they spotted sharks near the surface, they were always in motion. Therefore, they reasoned, sharks do not sleep. However, scientists have found that sharks do sleep, but their rest times are hardly akin to the eight or so hours of deep sleep that most people enjoy. Rather, for sharks, sleep is light and usually for no more than ten minutes or so at a time. And during that brief period of sleep, the sharks are usually in motion. Most sharks draw oxygen through their gills by moving forward. Thus to continue breathing, they must always be in motion—which they can accomplish even if they are asleep.

Since sharks lack bony skeletons, they are rather light for their size. Sharks are regarded as slightly negatively buoyant, meaning that if they remained still in the water they would sink—albeit very slowly. On the other hand, thanks to its very light body, it takes very little effort by a shark to move through the water, diving deeply to the bottom or propelling itself to the surface. Therefore, most species of sharks are constantly in motion—even when they are asleep.

Moreover, those sailors of yore might have thought the sharks were not asleep because their eyes were open. In fact, many sharks have been found to sleep with their eyes open. A 2022 study by marine biologists at the University of Western Australia found that different species of sharks have different sleep habits—including whether they close their eyes while sleeping. When the researchers studied sharks sleeping during the day, they found that most of them slept with their eyes closed. But for sharks that sleep at night, just 38 percent closed their eyes. The researchers speculated that most sharks close their eyes during the day because the brighter light, even beneath the ocean surface, makes it difficult for them to nap. However, many of the sharks sleeping at night found no reason to close their eyes. Said the study, "This suggests that eye closure is more likely associated with an external factor, such as the presence of light rather than sleep."[10]

The unusual habit of keeping their eyes open—even during sleep—illustrates why many species of sharks are recognized as some of the most unusual creatures on the planet. Sharks have outlasted the dinosaurs. Some sharks live for hundreds of years. Sharks enjoy a sixth sense, enabling them to detect electrical activity given off by nearby prey. These unique attributes are why many marine biologists have focused their research so keenly on sharks and why their research has uncovered so many fascinating habits and features found in the hundreds of species of sharks.

CHAPTER TWO

# Why Are Sharks Vital to the Environment?

Environmentalists liken the apex predator sharks to the lions of the Serengeti. The Serengeti is a large swath of land in the African nation of Tanzania. It sprawls over some 5,700 square miles (14,763 sq. km) of grassy plains and woodlands. These grassy plains and woodlands in the Serengeti—and similar regions elsewhere on the planet—are vital components in the international mission to reverse climate change.

Scientists around the globe agree that the climate of the earth is warming, due to the introduction of carbon dioxide into the atmosphere that commenced at the dawn of the Industrial Revolution in the late eighteenth century. The emissions from gasoline-powered automobiles, smoke from coal-fired power plants, and exhaust from the jet engines powering airliners are among the many sources that fill the atmosphere with the chemical carbon dioxide. Carbon dioxide is a so-called greenhouse gas, meaning it traps heat in the atmosphere, thus causing temperatures around the planet to rise.

The planet is already experiencing the dire consequences of the changing climate. This includes destructive storms, melting polar ice caps, rising ocean levels, an increasing number of droughts and wildfires, and the loss of many species of plants and animals. Scientists predict that these effects will worsen as climate change progresses.

Around the globe, political leaders hope to reduce the carbon content of the atmosphere through many methods. In Tanzania, the Serengeti has been declared a national park, protecting the region against the type of industrial development that fuels climate change. Grasses, trees, and other vegetation serve a role in fighting climate change: plants absorb carbon dioxide, thus drawing the greenhouse gas out of the atmosphere.

But the plant life that sprawls across the Serengeti must itself be protected. The Serengeti is heavily populated by gazelles, giraffes, antelopes, zebras, and other animals that feed on grass and leaves. If left on their own, these grazing animals would eventually reduce the Serengeti to a vast, dusty wasteland. And that is why the apex predator lions are so important to the Serengeti: by feeding on the grazing animals, they help keep those populations in check, which in turn enables the plant life in the Serengeti to thrive—thereby helping to slow climate change. "Nature needs lions to balance the wilderness by preying on grazers,"[11] says Edith Asiimwe, an Africa-based representative for the environmental group World Animal Protection.

## Blue Carbon Zones

Beneath the ocean surface are huge regions that resemble the Serengeti: undersea plains that are covered with seagrass and other vegetation. Seawater absorbs carbon dioxide from the atmosphere. And once those carbon dioxide molecules submerge into the ocean, they are drawn into the leaves, stems, and roots of underwater vegetation—effectively locking them away forever. These areas of undersea grasslands are known as blue carbon zones.

But there are many species of fish as well as amphibious animals that eat seagrass and other plants along the ocean floor. These fish and amphibious animals are known as herbivores. A single adult green turtle, which is usually found in tropical waters, can consume more than 4 pounds (1.8 kg) of seagrass per day.

A dugong—a marine mammal resembling a seal, found mostly in the shallow coastal waters around Australia—is known to consume about 88 pounds (40 kg) of seagrass per day. And a 2002 study by the Florida Keys National Marine Sanctuary looked at the grazing habits of a small colorful fish known as the parrotfish. Examining a region off the coast of Florida known as Hawk Channel, the study concluded that if left on their own, parrotfish would consume 85 percent of the seagrass in the channel.

Marine biologists estimate there are some 3.5 trillion fish and amphibious creatures living beneath the surface of the world's oceans that rely on seagrass and other vegetation for nourishment. If left on their own, they would eventually make the ocean floors barren of all plant life. Without these blue carbon zones, the carbon dioxide that falls into the sea would not stay below water for long—the greenhouse gas molecules would soon rise back to the ocean surface and bleed back into the atmosphere, further contributing to climate change.

*A green sea turtle feasts on seagrass in the Red Sea off the coast of Egypt. A single sea turtle can eat more than 4 pounds of seagrass per day.*

And that is why the apex predator sharks are so important to the health of undersea vegetation and, by extension, the climate of the planet. Says marine biologist John Long:

> Predator [sharks], like lions on the Serengeti, keep the herbivores in check . . . so the balance of nature doesn't get out of whack. If there were not lions, the antelopes and the gazelles would eat all the grasses and soon there would be no grasses and they would overpopulate and you would have a terrible ecological collapse. What sharks do in the oceans is similar. As top-line predators they keep [other] fishes in check.[12]

## Sharks Promote Biodiversity

As apex predator sharks feed on the herbivores, they are accomplishing more than just helping preserve seagrass and control climate change. Sharks promote biodiversity—an important aspect in the ecology of the world. Biodiversity ensures that numerous species will exist within various regions of the planet. A loss of biodiversity occurs when a specific species becomes dominant in a region. When such changes occur, it often means that other species of animal life will soon decline in that region and perhaps eventually become extinct.

> **"Predator [sharks], like lions on the Serengeti, keep the herbivores in check . . . so the balance of nature doesn't get out of whack."[12]**
>
> —John Long, marine biologist

Examples of extinctions caused by an overabundance of a dominant species can be found in the US territory of Guam, an island in the western Pacific Ocean. A tiny bird known as the bridled white-eye as well as a winged mammal, the Little Mariana fruit bat, are both believed to have become extinct during the past decade because of the growth and abundance of the predatory reptile known as the brown tree snake. Scientists believe that if an apex predator had been present in Guam, it would have controlled the population of the brown tree snake, and the bridled white-eye and Little Mariana fruit bat would not have gone extinct. The same can be said for other bird and reptile species. A report by the US Fish and Wildlife Service found that extinctions on the island have been occurring for many years because of the proliferation of the brown tree snake, which is a nonnative invasive species and thus it has no natural predators.

As with the bridled white-eye and Little Mariana fruit bat, many species of fish within the world's oceans could be threatened with extinction if they are continually hunted by predatory fish that are free to roam without apex predator sharks to keep their populations in check. Says Lawrence Chlebeck, a marine biologist at James Cook University in Sydney, Australia:

> Sharks are very good at promoting biodiversity, and . . . abundant biodiversity is really important to healthy ecosystems. Sharks keep food webs in balance, and encourage biodiversity to flourish by feeding on those species that are most numerous, allowing other species a chance to also grow in number. Good ecosystems need biodiversity, and in turn, sharks are an important element of healthy biodiversity.[13]

## Protecting the Coral Reefs

As Chlebeck notes, there are many predators living in the world's oceans, but not all of them are apex predators. Many species of fish, including several species of sharks, eat smaller fish for food—but are themselves prey for larger fish. These sharks, known as mesopredators, also play an important role in protecting the undersea environment.

> **"Good ecosystems need biodiversity, and in turn, sharks are an important element of healthy biodiversity."[13]**
>
> —Lawrence Chlebeck, marine biologist

The Caribbean reef shark, which typically grows to a length of about 9 feet (2.7 m), is a mesopredator. As its name suggests, the Caribbean reef shark can be found in the vicinity of the coral reefs of the Caribbean Sea, a portion

## Sharks Fins and Renewable Energy

Sharks offer benefits to humanity that go beyond healthy oceans. For the past two decades, researchers have experimented with creating energy from posts inserted into the seafloor. Attached to the posts are metallic fins designed to resemble the fins of sharks. As the ocean current rushes past the fins, the fins sway back and forth as though they are propelling a shark. This motion is then harnessed, creating electricity much the same way a wind turbine creates electricity by spinning in the wind. Researchers chose to model the devices after the fins of sharks because sharks are able to propel themselves forward with very little motion by their fins—meaning shark fins lend themselves well to creating energy.

By 2024 efforts to harness electricity through artificial shark fins were still in the trial phase. Researchers have not yet found a cost-effective way to establish the system. "Tides and river currents could be a huge boost to the US power grid," said Michael Lawson, a researcher for the National Renewable Energy Laboratory. "The resources are abundant and renewable. But compared to other energy generating sources, the technologies are still expensive."

Quoted in Caitlin McDermott-Murphy, "Getting in the Water with Sharks," National Renewable Energy Laboratory, May 31, 2022. www.nrel.gov.

of the Atlantic Ocean alongside Central America. Coral reefs are important components of the undersea environment. The reefs form when tiny boneless creatures known as corals form a colony. The corals discharge from their bodies the chemical calcium carbonate, which forms a hard, rocklike substance. As the colony grows, the calcium carbonate emitted from all the corals forms a hard, stonelike reef. Coral reefs play an important role in protecting shorelines from damaging storms and floods. The undersea reefs serve as barriers between ocean waves and the shorelines; by blocking the waves, the reefs reduce their impact on seaside communities. As the climate of the planet continues to warm, future storms become more fierce, and the ocean levels rise, coral reefs will play a vital role in protecting seaside communities.

Coral reefs are diverse ecosystems. They provide habitats for a wide array of ocean plants and animals. Many species of fish,

*The Caribbean reef shark, pictured here, is a mesopredator that preys on smaller animals, but itself is prey for larger sharks.*

among them parrotfish and butterfly fish, feed on the corals within the reefs. Moreover, the parrotfish is also known to actually chew away at the stonelike reef. If those fish are left unchecked, it is likely the reefs would eventually collapse. This would endanger other animals that depend on the coral reefs for survival. It would also expose shorelines to massive waves and flooding. But parrotfish are, in fact, kept in check by mesopredators that feed on them, such as the Caribbean reef shark. Therefore, the Caribbean reef shark as well as similar species of sharks play vital roles in protecting the coral reefs across the ocean floor.

Yet even these sharks and other mesopredators could become a threat to the ocean environment if their populations were to grow too large. The presence of apex predator sharks ensures that this does not happen. The apex predators help control the populations of various species, often ensuring that no single species of fish or amphibious creature becomes dominant in a particular region of the undersea world. They help control these populations by preying on the mesopredator fish they encounter. In addition, an apex predator's mere presence in a region of the sea often sends a message to the mesopredator fish that they would do well to leave and go elsewhere. Researchers at Florida International University report that bigger sharks are very good at chasing away the mesopredators and herbivores, forcing them into new territories. According to a report from the university, "The largest sharks of some of the biggest species play an oversized role in healthy oceans. Sometimes their sheer size is enough to scare away prey"[14] that could severely damage coral reefs and other undersea natural resources.

Sharks are therefore greatly responsible for the constant movements of different species of fish from one region of the ocean to another. It is a function that helps maintain biodiversity throughout the oceans of the world.

## Essential Nutrients for Ocean Plants

Sharks do more than just help ensure that numerous species of aquatic life are constantly on the move, promoting the spread of biodiversity. Sharks are also responsible for helping in the transfer of nutrients from region to region beneath the waves. Nutrients are natural chemicals, such as nitrogen, phosphorus and potassium, that are vital to maintaining healthy animals and plants. Nitrogen, for example, enhances the growth potential of the typical green plant. Plants that are lacking in nitrogen are often pale, with fragile leaves and stems. They do not grow as fully or for as long as plants with healthy nitrogen content. Similarly, animals—including humans—rely on nitrogen for heathy growth and to fuel immunity against diseases.

Sharks are known to swim thousands of miles during their lifetime—and since sharks have an incredibly long lifetime, it is easy to see how they have become such accomplished world travelers. As they travel the world beneath the sea, they are

## Sharks Help Keep Hospitals Germ-Free

Sharks live very long lives, a fact scientists attribute in part to the surface of the shark's body. Sharkskin is composed of myriads of tiny V-shaped scales known as denticles. The denticle patterns across the shark's body are very resistant to germs: the germs do not adhere well to the patterns. Therefore, sharks do not get sick very often, if at all.

Researchers have studied whether artificial surfaces could be created to resemble denticle patterns in sharkskin. Those surfaces are then employed in hospitals, which due to the ill and contagious patients under their care, face challenges in keeping their facilities sanitary. The adhesive surfaces are applied to doorknobs, faucets, restroom doors, and other places that are touched by patients, staff members, and visitors—preventing them from spreading germs through hand contact.

In 2024 Aurora, Colorado–based Sharklet Technologies developed shark-skin-like gloves to be worn by physicians as they examine their patients. Says Dr. Margaret Sande, a professor at the University of Colorado School of Medicine, "We find that hand contact—as clinicians [we] are going quickly from room to room—is often a means of transmission. So, the hand can be sort of the source of all evil."

Quoted in Sharklet, "*PBS NewsHour*: Sharklet Is Expected to Become a Key to Resist Superbugs," August 13, 2024. www.sharklet.com.

constantly eating and, of course, leaving their waste matter behind. This waste matter contains nutrients—which helps maintain healthy ecosystems.

Although humans are quick to flush their waste matter down the toilet, it does have positive uses. The fertilizers used by farmers and home gardeners typically contain generous amounts of manure—largely provided by cattle, sheep, and chickens. Fertilizer is effective in nurturing plant life because it contains many nutrients that enhance the growth of vegetation.

Seagrass and other undersea plants also need fertilizer to help them grow. Those plants benefit from the nutrient-rich waste matter of sharks that constantly falls to the seafloor. "They recirculate nutrients from one zone of the ocean to another," says John Long. "You will see that top predators in the top waters are eating prey there but then they carry certain nutrients down to the lower levels or other parts of the ocean where they are needed."[15] And when there is a generous supply of nutrients beneath the waves, other aquatic life is sure to benefit.

## The Benefits of Vertical Swimming

Moreover, sharks are known to be vertical swimmers. Although they do swim near the water's surface, sharks typically also do a lot of diving. In other words, sharks are constantly going up and down, traveling from the waters near the surface to the ocean floor below. And as they make these constant journeys up and down in the ocean, their bodies pick up and spread nutrients that are floating in the water. Their bodies attract nutrients found near the ocean floor, then transfer them to regions near the surface. This is a particularly important function that aids microscopic forms of plant life known as phytoplankton, which can be found near the ocean's surface. The phytoplankton take in the deepwater nutrients spread by sharks' bodies. Aside from being a source

*Sharks carry nutrients from deep regions of the ocean to surface waters, providing food for the microscopic plants called phytoplankton (shown).*

of food for many ocean creatures, phytoplankton is also an important weapon in the fight against climate change. Says Alexander Nicolas, who helps lead programs for the Switzerland-based conservation and environmental organization the World Wildlife Fund:

> Sharks swim throughout the different ocean layers, performing vertical migrations that mix nutrient- and oxygen-rich deep water with nutrient-poor surface waters, helping to oxygenate them and enhance their productivity. This aids creatures living in the ocean's surface water, like phytoplankton, by getting the nutrients they need to be more productive. Like plants on land, phytoplankton can pull carbon dioxide from the atmosphere and produce oxygen.[16]

**"Sharks swim throughout the different ocean layers, performing vertical migrations that mix nutrient- and oxygen-rich deep water with nutrient-poor surface waters, helping to oxygenate them and enhance their productivity."[16]**

—Alexander Nicolas, World Wildlife Fund conservationist

Sharks play a vital role in the maintenance of the world's ocean ecosystems. Beneath the seas, the apex predator and mesopredator sharks help promote biodiversity by chasing other fish into new territories. And when they chase other predators away, sharks help ensure that many species of sea life will not go extinct. Moreover, sharks act as protectors of the coral reefs, attacking the fish that prey on the corals. They are also vital allies for humans and other creatures on land. By preying on herbivores, sharks help preserve the huge beds of seagrasses beneath the waves. And by protecting those beds of underwater vegetation, sharks are helping stave off the devastating effects of climate change.

CHAPTER THREE

# What Are the Threats to Sharks?

Steven Spielberg is one of Hollywood's most influential and successful filmmakers. Over the years, Spielberg has served as director, producer, and writer for numerous films that have earned hundreds of millions of dollars at the box office and remain very much in demand for TV network broadcasts and streaming services. Moreover, his films have won numerous Academy Awards as well as other honors. When Spielberg's movies are released to the public, critics generally rave about his films. "No director or producer has ever put together a more popular body of work,"[17] wrote noted film critic Roger Ebert.

Spielberg's first big hit was released in the summer of 1975. The film told the story of a small American seaside community terrorized by a great white shark roaming offshore. The shark attacks and violently kills swimmers, sport fishers, and others. The title of the film is, simply, *Jaws*. (The movie was based on a novel of the same name, written by author Peter Benchley and first published in 1974.) The film was a blockbuster, earning more than $260 million at the box office. And in many ways *Jaws* changed filmmaking as directors learned that the type of shocking scenes Spielberg included in the film could sell a lot of tickets. For example, the film opens when a young woman, out for a romantic dip in the ocean with her boyfriend, is suddenly mauled by the villain of the film—a great white shark.

> "That's one of the things I still fear—not to get eaten by a shark, but that sharks are somehow mad at me for the feeding frenzy of crazy sport fishermen that happened after 1975."[18]
>
> —Steven Spielberg, film director

In 2022, nearly five decades after *Jaws* premiered in American movie theaters, Spielberg told a journalist that he regretted ever making the movie. Spielberg blamed the film for helping create fear and hatred of sharks. The portrayal of sharks as sinister, relentless predators, he believes, sparked a long-term assault on the world's shark populations. That assault, he says, has been carried out by people involved in sport and commercial fishing, with the result that several species of sharks are now either threatened or endangered. "That's one of the things I still fear—not to get eaten by a shark, but that sharks are somehow mad at me for the feeding frenzy of crazy sport fishermen that happened after 1975," Spielberg said. "I truly, and to this day, regret the decimation of the shark population because of the book and the film."[18]

## A New Perception of Sharks

Spielberg's belief that his film sparked a worldwide warfare on sharks is shared by others. Says Berkeley, California–based marine biologist David McGuire:

> It's significant to note that the negative portrayal of sharks in *Jaws* has had real-world consequences. Sports fishermen catch and display sharks without regard to cruelty. Coastal ocean-goers ignorant about true shark behavior and relative risk are overly fearful of sharks, and this perception can generate overly-reactive public policy in the event of a (rare) attack on a human. The general attitude about sharks has led from a general disregard and mismanagement of sharks . . . to an outright vilification of these apex predators.
>
> Additionally, the negative perception of sharks has made it difficult for conservation efforts to gain traction in some areas, as many people are reluctant to support measures to protect these important animals.[19]

In fact, the desire by sport fishers to reel in sharks exploded within weeks of the film's release in 1975. In the seaside community of Montauk on Long Island in New York State, business boomed that summer and fall for chartered fishing vessels, with customers demanding that their excursions specifically target sharks. "That movie broke the [shark fishing] business wide open," said Glenn Frankel, captain of the chartered fishing vessel *Blue Water*, which sailed out of Montauk. "I would say we had a 50 percent increase this season of people chartering boats to go out after shark and other big game."[20] Another skipper, Frank Mundus, said, "When I

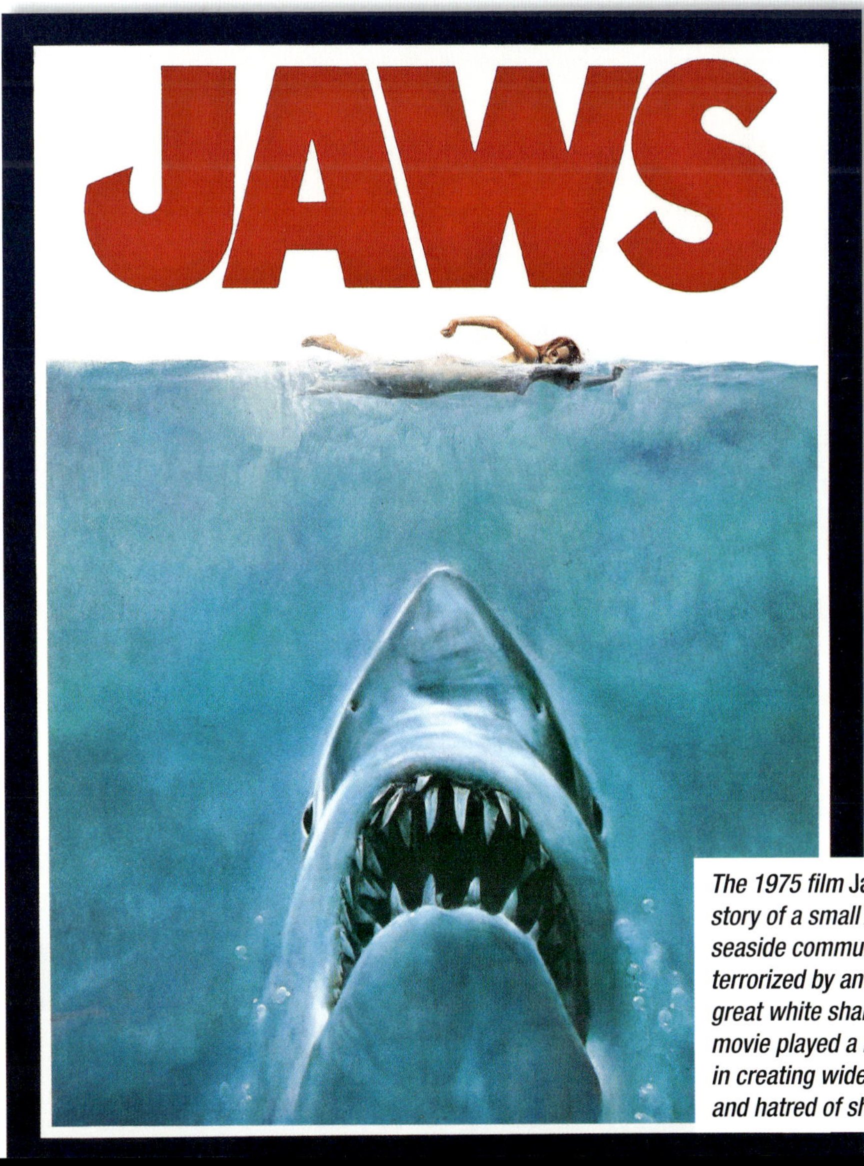


*The 1975 film* Jaws *told the story of a small American seaside community terrorized by an enormous great white shark. This movie played a huge role in creating widespread fear and hatred of sharks.*

came here some 10 years back, everyone used to turn their nose up on shark. They called it a poor man's big fish; now everyone's an expert and capitalizing on the movie."[21] In fact, that summer a local business in Montauk sponsored a shark-fishing competition: some six hundred fishers entered the contest. The winner reeled in a 399-pound (181 kg) dusky shark—a black-skinned predator that typically grows to about 13 feet (4 m) in length.

## Threatened and Endangered

Over the decades, the slaughter inflicted on sharks by sport and commercial fishing has taken its toll on numerous species of sharks. According to the National Oceanic and Atmospheric Administration (NOAA), an agency of the US government, the annual loss of some 80 million sharks to sport and commercial fishing has resulted in twenty-one species of sharks being placed on either the endangered or threatened species lists. Under the 1973 Endangered Species Act, endangered species are species of animals that are in immediate danger of becoming extinct; threatened species are species of animals that are in danger of facing extinction in the near future. Some sharks that can be found near US coasts have been placed on the threatened and endangered lists. For example, a species found on NOAA's threatened list is the scalloped hammerhead shark, a sleek, gray-colored shark that grows to a length of about 13 feet (4 m) and can be found off the West Coast. Meanwhile, the oceanic whitetip shark, a gray shark that grows to a length of about 11 feet (3.4 m), has been placed on the endangered list. This shark can be found swimming off the coast of New England.

In addition to the threats sharks face from commercial and sport fishing, sharks are slow to repopulate their species due to a number of other factors. Depending on the species, many female sharks carry their young for one to three years. These long periods mean they do not reproduce quickly. The predatory nature of sharks toward their siblings—those baby sharks

## The Fear of Sharks

Films like *Jaws* and news stories on fatal shark attacks can spark a condition known as galeophobia. The term has its roots in the Greek words for "sharks" (*galeos*) and "fear" (*phobos*).

Tyler Bowling, manager of the Florida Program for Shark Research at the Florida Museum of Natural History, likens galeophobia to the fear many people feel as they walk through a forest at night. Pacing through darkness, they feel a chill in the air while hearing threatening sounds from animals they cannot see. In reality, Bowling says, the animals are mostly warning other animals to flee the area where the human is walking. He says:

> The ocean is kind of like the woods at night. . . . The marine world can be strange and overwhelming at times. Our mind creates monsters where there are none. . . . Symptoms [of galeophobia] may include anxiety with elevated heart rate, shortness of breath, shaking, sweating, nausea, or dizziness. Hollywood movies and the media reinforce the fear of sharks. No matter the level of injury, many news agencies post a picture of a huge white shark with a gruesome headline. Many of the bites we see here in Florida are from sharks that were smaller than the person they bit.

Tyler Bowling, "The Fear of Sharks," Florida Museum of Natural History, January 19, 2022. www.floridamuseum.ufl.edu.

that are lucky enough to hatch from their eggs first often eat their mother's unhatched eggs—affects population numbers as well. "The sharks that are especially slow-growing, sought-after and unprotected from overfishing tend to be the most threatened,"[22] says marine biologist Nicholas Dulvy, who led a 2019 study of endangered shark species for the Swiss-based environmental advocacy group International Union of Conservation of Nature.

### Victims of Bycatch

As popular as shark fishing has become among sport anglers, it is commercial fishing that has contributed most to the decline in the global shark population. According to a 2015 report by the

Food and Agriculture Organization of the United Nations, shark products—flesh, fins, and oil from shark liver—earn more than $1 billion a year in sales. Many sharks are caught for commercial sales the same way sport fishers catch them—through baited hooks at the ends of fishing lines. Many are also snagged in nets dragged through the ocean by commercial fishing vessels. And in most cases the vessels snagging those sharks are not specifically fishing for sharks. They are seeking other species of fish with the intent of making them available to the seafood market. But the sharks get hauled in with the rest of the fish and are then chopped up and sold along with the tuna, halibut, salmon, and cod. Sharks are thus victims of what is known as bycatch. They are not the intended prey of the commercial fishing vessels, but the sailors aboard those vessels do not free them from their nets. Says marine biologist John Long:

> Sharks have mostly been threatened due to overfishing. It's not just that people are out there fishing for sharks . . . you get these giant ocean trawlers that are catching the edible fishes that we like to eat. The sharks are caught as a bycatch and killed. There are many hundreds of tons of sharks killed that way each year. It isn't regulated and we don't know the numbers of species that have been wiped out that way.[23]

According to NOAA, the scalloped hammerhead shark and the oceanic whitetip shark are both victims of bycatch. Many of the other threatened and endangered sharks on the NOAA lists, typically found in foreign waters, are victims of bycatch as well.

Even though sharks are not always the intended targets of commercial fishers, there are nevertheless many consumers throughout the world who enjoy dining on different dishes composed of sharks. One of the most popular products drawn from sharks is soup made from shark fins. Centuries ago, shark fin soup was very popular in Asian countries, but the taste for the

*Commercial fishing vessels often scoop up sharks by accident when trawling for a different catch, such as shrimp or fish.*

soup died out because of the steep expense of catching the sharks, slaughtering the animals, and creating soup from the fins. Starting in the 1980s, though, the demand for shark fin soup exploded in China. The country's economic growth during the era meant many citizens could now enjoy luxuries previously out of their reach, and one of those luxuries was shark fin soup. Moreover, the taste for the soup eventually found its way to other markets. According to a report prepared by the California-based Professional Association of Diving Instructors Foundation, in addition to China there are major demands for shark fins in countries as diverse as Canada, Indonesia, Peru, South Africa, United Arab Emirates, and the United States.

When asked by a journalist to describe the taste of shark fin soup, Alvin Leung, owner of Bo Innovation, a popular restaurant in the Chinese city of Hong Kong, said simply, "It's like champagne."[24] *Time* magazine reporter Krista Mahr had a decidedly different opinion of shark fin soup. She wrote, "A

> "I knew [shark fin] soup had a whiff of controversy around it, but I hadn't yet formed a personal policy, so I gave it a try. I found it underwhelming."[25]
>
> —Krista Mahr, journalist

server came to our table with a cluster of small white bowls, which a few of my colleagues politely declined. I knew the soup had a whiff of controversy around it, but I hadn't yet formed a personal policy, so I gave it a try. I found it underwhelming."[25] Nevertheless, people who enjoy shark fin soup tend to agree more with Leung than with Mahr. According to the Princeton, New Jersey–based environmental group the Shark Research Institute, at least some 73 million of the 80 million sharks killed each year are killed exclusively for their fins.

## Wallets, Belts, Shoes, and Cosmetics

Sharks do more than just provide seafood to consumers. Over the years, entrepreneurs have found other uses for sharks caught by commercial fishers. Sharkskin has been employed as a substitute for ordinary leather, which is typically tanned from the hides of cattle, goats, pigs, and sheep. Wallets, belts, shoes, and similar goods made of sharkskin are very much in demand—and often very expensive. A belt made from sharkskin is likely to cost several hundred dollars, while an ordinary cowskin leather belt can usually be purchased for twenty dollars or so.

Therefore, sharkskin products are often purchased by wealthy shoppers—but advocates of sharkskin products recommend that shoppers with tight budgets would do well to give sharkskin a try. Karl Wasson, a Vietnam-based expert on shoe styles, says, "Shark hide is eye-catching. It's visually distinct from most terrestrial animal hides. In place of pores, there are scales. Because most tanneries grind down the . . . scales, shark hide has a unique and lovely deep pebble texture."[26] However, as is the case with sharkskin belts and wallets, sharkskin shoes can be very expensive. The California-based company Vaquero Boots specializes in selling cowboy boots made from sharkskin. The typical price for a single pair of the company's sharkskin cowboy boots is more than $300. A pair of cowboy boots made from ordinary leather typically costs about $70.

The commercial potential of sharks goes well beyond their use to manufacture finely textured cowboy boots and other expensive consumer goods. The cosmetics industry also relies on commercial shark fishing. A substance known as squalene is an important ingredient in many products employed to treat dry skin. Squalene helps hydrate the skin. It has also been found to be effective in reducing inflammation, especially soreness that may linger after a minor injury. Squalene is refined from oil drawn from the livers of sharks.

Squalene is also a common ingredient found in cosmetics, including lipstick, foundation, eyeshadow, moisturizer, sunscreen, and lip balm. The substance has numerous effects that make it an important ingredient of cosmetics: squalene is nongreasy, thus

## Shark Fin Soup and Mercury

The Shark Research Institute estimates that some 73 million sharks are killed each year specifically to provide fins for shark fin soup. Many people who enjoy the delicacy do not realize that shark fin soup is very unhealthy. A 2020 study by Florida International University found that shark fins contain high levels of mercury.

Concentrations of the element mercury cause damage to human organs. According to the study, sharks accumulate mercury in their bodies through the discharge of sewage into the ocean and through the burning of oil and coal. When those fuels are burned, mercury vapors enter the atmosphere, but the mercury eventually drops into the ocean and is absorbed by sea life, including sharks.

When mercury enters a shark's body, it accumulates in the fins and remains in the fins after they are cut off and cooked in soup. Says Gary Stokes, director of the Hong Kong–based environmental advocacy group Oceans Asia, "The average person has no idea of the high mercury levels in shark fins. In fact, very few have any idea of where the dish in front of them came from at all. . . . Many don't even connect that it came from an actual shark."

Quoted in Elizabeth Claire Alberts, "Mercury with That? Shark Fins Served with Illegal Doses of Heavy Metals," Mongabay, June 19, 2020. https://news.Mongabay.com.

enhancing the spreadability of cosmetics, and it enables cosmetics to stay on the skin longer. Says Lauren Evashenk, founder of Naked Truth Beauty, a Washington State company that markets cosmetics made completely with ingredients drawn from plants:

> One of the most prominent ingredients in cosmetics is responsible for fueling a multimillion-dollar industry with deadly consequences. Squalene, an oil compound found in shark livers, is highly regarded for its moisturizing and restorative properties and has been finding its way into everyday cosmetics for years. But at what cost? 2.7 million sharks are captured and killed for their livers each year. The primary culprit behind this inhumane hunt driving the extinction of endangered and vulnerable deep-sea species? The cosmetics industry.[27]

## The Thrill of the Catch

While commercial fishing vessels serve the needs of diners and the cosmetics and fashion industries, sport fishing remains a significant threat to numerous species of sharks. In the decades since the release of *Jaws*, the sport of shark fishing has virtually never seen a downturn in popularity. Sport fishers constantly flock to seaside communities with the desire to reel in sharks—and then pose for dockside photos with their catches dangling beside them from a rope.

Around the globe, associations of sport fishers have been established to sponsor tournaments and provide news about the latest gear and techniques that can be employed to reel in sharks. Among these groups are the Jersey Coast Shark Anglers and South Florida Shark Club in the United States and, across the Atlantic, the Shark Angling Club of Great Britain. On social media, numerous groups on Facebook, Instagram, and X are devoted to the sport of shark fishing. Many of these groups have thousands of followers. On Facebook, for example, the group Land Based Shark

*A blue shark is weighed after being caught at a shark fishing tournament. Sport fishing is a threat to numerous shark species.*

Fishing World Wide has more than four thousand followers. Members post updates and news in the world of shark fishing several times a week. The Facebook group's cover photo depicts a sport angler prying open the mouth of a dead hammerhead shark.

Moreover, seaside communities often serve as bases for the owners of vessels that are chartered for sport shark fishing outings. Many other sport fishers prefer to reel in sharks from beaches or piers. "I caught my first shark when I was six or seven years

> "I caught my first shark when I was six or seven years old. It almost pulled me off the pier. Ever since then I've been hooked on shark fishing from piers and the beach."[28]
>
> —Adam Lopez, sport fisher

old," says Adam Lopez, a dedicated shark angler from Bradenton, Florida. "It almost pulled me off the pier. Ever since then I've been hooked on shark fishing from piers and the beach."[28]

Lopez recalls his most impressive catch in 2022 while fishing with some friends from a beach in Bradenton. The group had already snared a mako shark. A short time later, Lopez says, "We hooked another big shark and this one ran off 600 to 700 yards. . . . We could tell it was jumping and spinning with how the line was acting but it was dark by then so we couldn't see it." After about forty-five minutes of wrestling with his line, Lopez managed to draw the shark near the beach. "To all our surprises, I had landed a juvenile great white shark. Never in my wildest dreams would I have thought to land a Florida great white shark from the beach. I've only heard of two other people doing that, and no one ever landed a great white and a mako [shark] in the same night."[29]

Lopez's comments illustrate that hunting for sharks can be exciting and breathtaking adventures. And that is why a robust interest in sport shark fishing remains a part of modern culture. Meanwhile, the taste for shark fin soup, the use of sharkskin to manufacture consumer products like belts and boots, and the use of shark oil by the cosmetics industry has resulted in the loss of tens of millions of sharks per year from the world's oceans. Continued high demand for shark-related products could push several shark species toward extinction.

## CHAPTER FOUR

# What Can Be Done to Save Sharks?

Although many countries have banned commercial fishing that targets sharks for their fins, the shark fin trade has not abated. In 2022 the US Congress passed the Shark Fin Sales Elimination Act, making the sale, possession, transport, and purchase of shark fins and products containing shark fins—including shark fin soup—illegal throughout the country. In addition, the states of California, Delaware, Florida, Hawaii, Illinois, Maryland, Massachusetts, Nevada, New York, Oregon, Rhode Island, Texas, and Washington as well as the US territories of American Samoa, Guam, and the Northern Mariana Islands have enacted their own laws prohibiting the trade in shark fins.

Despite these laws, in 2023 NOAA agents seized some 1,400 pounds (635 kg) of dried shark fins from a Kentucky-based import company. The value of the fins—the price the importer hoped to earn from US restaurants planning to buy the fins—was estimated at more than $75,000. Said Gib Brogan, an activist with the Washington, DC–based environmental group Oceana International, "We've seen this first case, but we're hoping that [it's] going to be one in a string so that we can finally get the US out of the shark fin trade."[30]

Traders in shark fins have also been charged under other laws. For example, in 2018 a commercial fishing vessel docked in Hawaii was found to contain some 200 pounds (91 kg) of shark fins. The owner of the vessel, a company based in Japan, was criminally charged under the Lacey

> "Shark finning is inhumane, intolerable, and takes a very real toll on our precious ocean ecosystem."[31]
>
> —Kenji M. Price, federal prosecutor

Act, a US law that makes it illegal to hunt and kill threatened or endangered species. Prosecutors alleged that the fins had been removed from shark species that are on the NOAA threatened and endangered species lists. Executives from the company pleaded guilty to the violation, agreeing to a pay a fine of $126,000 and forfeit the fishing vessel, valued at $119,000, to federal authorities. Japanese authorities also penalized the company, forcing it to surrender its license to fish in Japanese waters. Said Kenji M. Price, a federal prosecutor in Hawaii who brought the criminal charges against the Japanese seafood company, "Shark finning is inhumane, intolerable, and takes a very real toll on our precious ocean ecosystem. My office is committed to combatting this cruel practice by prosecuting to the fullest extent of the law anyone found to be trafficking in these types of shark fins."[31]

## Available in American Restaurants

As these cases illustrate, there have been criminal prosecutions brought against traders in shark fins, but environmentalists believe many of the federal and state laws designed to protect sharks are poorly enforced or do not go far enough to ensure protection for sharks. For example, the Washington, DC–based environmental group Animal Welfare Institute (AWI) reports that numerous restaurants in the United States continue to sell shark fin soup. The organization even maintains a website that reports the names and addresses of restaurants serving the soup. Its list is developed from reports made by volunteers.

By 2024 the list included hundreds of restaurants in thirty-two states as well as three US territories. In California, for example, volunteers reported finding shark fin soup on the menu in twenty-nine restaurants; in Indiana a single restaurant in the city of Indianapolis was cited. AWI pointed out that in the eighteen states where no restaurants were reported, there could very well be restaurants featuring shark fin soup on their menu. The organization contends that there are restaurants likely serving shark fin soup in many American communities where AWI lacks local volunteers willing to search for them

Moreover, many environmentalists insist that the Shark Fin Sales Elimination Act, although well intentioned, has done little to protect sharks. The law prohibits capturing and slaughtering sharks specifically for their fins. But a 2024 study by marine biologists at Dalhousie University in Halifax, Canada, found that commercial fishers are continuing to catch as many sharks as they were before the enactment of the law. Only now, the report suggests, they have stepped up efforts to sell the shark meat as well as the fins—meaning that they are in compliance with the law because they are now doing more than selling only the shark fins to restaurants and other commercial buyers of food products. "We should have seen a signal in reduced mortality [following adoption of the law]," says Boris Worm, one of the authors of the study. "The surprising result is that we didn't."[32]

*Dried shark fins are displayed for sale in Hong Kong. The fins are the main ingredient in shark fin soup, which is a popular traditional dish in some parts of the world. More than 70 million sharks are killed each year for their fins.*

### Protecting Hammerhead Sharks in Costa Rica

The 1973 Endangered Species Act made it illegal to capture or kill for commercial or sport purposes any animals included under NOAA's list of threatened and endangered species. Other countries have adopted similar laws. The government of Canada established the Species at Risk Act in 2002. Lawmakers in Great Britain adopted the nation's Wildlife and Countryside Act in 1981. Many other countries have passed similar laws—all aiming to protect creatures that its scientists have deemed to be among threatened and endangered species.

Some countries have only recently agreed to protect sharks judged to be among threatened and endangered species. For example, sharks that fall under several hammerhead species have been designated by NOAA as threatened or endangered, meaning it is illegal to fish for the sharks in US waters. But for many years, no such protection existed in the nation of Costa Rica—a Central American country located some 2,400 miles (3,862 km) from the southern US border. Hammerhead sharks have been found in abundance in the region. Their presence has proved popular among sport and commercial fishers, who flocked to Costa Rican beaches with the mission to reel in hammerheads.

Over the years, the Costa Rica–based environmental group Center for Rescued and Endangered Marine Species (CREMA) made appeals to the Costa Rican government demanding that lawmakers in that country prohibit sport and commercial fishing for hammerhead sharks as well as other shark species that may be imperiled. CREMA found allies in the Costa Rican Legislative Assembly who introduced many bills over the years to protect hammerhead sharks. Finally, in 2023 Costa Rican president Rodrigo Chaves Robles signed a law prohibiting sport and commercial fishing for hammerhead sharks.

Environmentalists welcomed the new law, but critics complained that the Costa Rican government acted too late, citing evidence that some 90 percent of the hammerhead shark population in Costa Rican waters had been lost by the time the new law

*Hammerhead sharks are abundant in the waters off Costa Rica. Sport and commercial fishing drastically reduced the area's shark population in the early 2000s, but laws now protect these creatures.*

was signed. "This is too late and it's way too little," says marine biologist Randall Arauz, who points out that environmentalists first called the government's attention to the loss in the hammerhead shark population back in 2013—ten years before the law was finally adopted. Says Arauz, "It was the [government's] obligation to ban hammerhead shark fishing commercialization, extraction, everything. That's been our battle for the last 10 years."[33]

## Shark Preservation an Emerging Priority

Still, CREMA's campaign to protect the hammerhead sharks swimming off the coast of Costa Rica illustrates how saving sharks has emerged as a priority for environmentalists. And in the United States as well as many other countries, shark protection is clearly a priority for environmental activists who believe more pressure must be exerted on elected officials to protect sharks.

The campaign to eliminate shark fin soup from people's diets has taken some steps forward. For example, in 2019 activists and municipal leaders in Hong Kong led a campaign to remove the soup from hotel menus. Ultimately, forty-seven hotels agreed to remove the soup from their menus. Still, activists agree that much work remains to be done. Reported the Hong Kong–based environmental group Bloom Association, "Shark fin–related products remain easily available in the local market. This is not too surprising, as Hong Kong is nonetheless conventionally one of the key consumption markets and trade hubs for shark fin–related products."[34]

As in Hong Kong, many private companies that operate hotels, resorts, and similar tourist destinations around the globe have responded to pressure by activists to ban shark fin soup. For example, hotels owned by corporations such as Hilton, Marriott, Ritz-Carlton, and Sheraton—which operate hundreds of inns and resorts—have removed shark fin soup from their menus in dozens of countries. Moreover, the organization known as Fin Free Thailand has persuaded more than one hundred hotels, restaurants, and supermarkets doing business in the Asian nation to no longer sell shark fin soup.

Activists are also working to expand the law prohibiting trade in shark fins to other products as well—such as wallets, belts, and shoes made of sharkskin as well as cosmetic products con-

## Progress in China

The government of China has taken what critics regard as a minor step to protect sharks. In 2013 Chinese lawmakers adopted a measure banning shark fin soup from being served at all official government banquets, receptions, and similar events. Nevertheless, the sale of shark fin soup to Chinese consumers remains legal throughout Asia's most populous nation.

Still, in recent years shark fin soup sales have been on the decline in China, largely because environmental activists have called attention to the effect on shark species. One of those activists is former professional basketball player Yao Ming, who is regarded as one of the nation's top sports stars. In a filmed 2009 interview, Yao pledged to give up shark fin soup, and he urged others in China to follow his example. He said, "When the buying stops, the killing can too."

Quoted in John Vidal, "This Could Be the Year We Start to Save, Not Slaughter, the Shark," *The Guardian* (Manchester, UK), January 11, 2014. www.theguardian.com.

taining shark squalene. For example, activists in Austria found success in 2022 when the Austrian Parliament adopted a law banning the import of virtually any product that is drawn from sharks—including food, cosmetics, and sharkskin wares. Austria is a landlocked country in Europe; therefore, all seafood and other products drawn from sea creatures must be transported into the country. “This also affects us Austrians, as every second breath we take comes from the ocean,”[35] says Lukas Hammer, an Austrian lawmaker who supported the act. Hammer notes that prior to the ban, Austria was one of Europe’s leading importers of shark-related products.

## Protecting Sharks and People

Efforts by environmentalists to ban the import of products made from sharkskin have not yet succeeded in the United States, but some states have adopted laws that they hope will severely cut down on the threats of extinction faced by many species of sharks. In California, for example, Governor Gavin Newsom signed a law in 2022 designed to protect great white sharks from sport fishing. The law makes it illegal to fish for great white sharks within 1 mile (1.6 km) of beaches and piers in California.

The law means that sport fishers can no longer simply walk onto California beaches or piers, cast their lines into the water, and wait for great white sharks to bite. Rather, to hunt for great white sharks, sport fishers are now forced to charter boats to take them beyond the 1-mile barrier—which is likely to cost nearly $1,000 dollars for a single excursion. The intent of the new law is to expose fewer great white sharks to sport fishing.

Also, the new law is seen as a protection for swimmers, surfers, and others who enjoy California beaches. Great white sharks are apex predators, and there have been attacks reported on humans by sharks that were clearly drawn to shallow waters by baited fishing lines. Therefore, the law’s intent is to expose fewer swimmers and others to attacks by great white sharks simply

> "Sharks are one of California's most iconic marine species, and it's our responsibility to ensure that their populations are sustained."[36]
>
> —Steve Bennett, California State Assembly member

because those sharks would not be lured to shallow waters by fishing lines. "Sharks are one of California's most iconic marine species, and it's our responsibility to ensure that their populations are sustained," says Steve Bennett, a member of the California State Assembly who sponsored the law. "At the same time, public safety is of the utmost importance."[36]

While environmental activists have won new laws in Costa Rica and Austria as well as in California and other state governments in the United States, there are nevertheless many countries that have refused to take steps to protect sharks. Some foreign governments recognize that their national economies rely greatly on shark tourism—in which sport fishers travel to their shores to fish for sharks that they would not be able to hunt in the United States or other countries. And many foreign governments recognize that business would dry up for commercial fishers in their country if they were restricted by law from catching sharks. According to Oceana International, countries having the fewest laws regulating shark fishing are Bolivia, Indonesia, Iran, Portugal, Spain, and Venezuela. Indeed, according to Oceana International, 67 percent of the fish caught by Spain's commercial fishing vessels consists of sharks, totaling some 41,200 tons (37,376 metric tons) of sharks per year.

## Learning Lessons from Sharks

Environmentalists, marine biologists, and others concerned with the welfare of sharks continue to work for protective laws, but many scientists believe a change in thinking about sharks is desperately needed. As shark populations continue to decline, and as additional species are added to the endangered and threatened lists, much valuable information about sharks may soon be lost. Scientists point out that shark species have survived some 400 million years—outlasting dinosaurs and many other animal

*A large, dried sharkskin is displayed in China. Dried sharkskin can be used to make boots, handbags, and other products. Environmentalists are working to ban the import of sharkskin into the United States, but they have not yet succeeded.*

species. Therefore, sharks have proved themselves to be expert at adapting to new and changing environments.

With citizens of the earth facing potentially devastating environmental consequences due to climate change, scientists argue that instead of killing sharks, humans should be devoting more efforts to studying them. "We have a lot we can learn from them," says marine biologist Jasmin Graham. "They have lived through Ice Ages, they have lived through lots of changes in the environment. Our environment is changing. There is a lot we can learn from sharks."[37]

A.J. Rotondella has gotten the message. Rotondella started fishing as a young boy. In 2011, while fishing from a beach in New Jersey, he caught his first shark. Rotondella was hooked, returning to the beach many times to continue his passion for shark fishing. As an adult, he found a job as a crew member on a boat

## A New Shark Tourism

For decades, shark tourism consisted mainly of sport anglers who traveled to beachfront communities, where they chartered boats for fishing expeditions or instead fished for sharks from beaches and piers. In recent years, though, activists and entrepreneurs have worked to establish a new form of shark tourism: inviting visitors to take boat rides into ocean waters, where they can witness sharks in their natural habitat—and perhaps come to recognize sharks as species worth preserving.

Kris Roszack leads such tours aboard his boat, which sails out of the beachfront town of Harwich on Cape Cod in Massachusetts. Marine biologists have determined that great white sharks, once few in number in the Cape Cod region, have enjoyed a resurgence in recent years. Scientists suggested that a state law protecting seals adopted back in 1972 helped spark growth in the seal population—which in turn drew many apex predator sharks to the region, particularly great white sharks.

And that is why Roszack tells his passengers they are virtually guaranteed to see great white sharks as the boat cruises through the Cape Cod waters. "I mean, success rate is probably in the upper 90 percent if the conditions are right," Roszack says.

Quoted in Paul Marino and Susanna Capelouto, "Pricey Great White Shark Tours Take Off in Cape Cod," National Public Radio, August 3, 2024. www.npr.org.

leased to host weddings and other parties. But he soon quit that job so he could devote himself full time to catching sharks.

Today, Rotondella is the proprietor of Apex Anglers, which leads shark-fishing outings for clients in New Jersey and Florida. Rotondella and his clients fish from the beaches only. He says, "To catch a fish with your feet on the sand without being able to chase them down, I find that to be a lot more fun. And also, when you catch a huge animal like a shark you get right up next to it instead of looking over a [ship's] bow at it."[38]

But Rotondella is also very aware that the worldwide shark population has been severely impacted by sport and commercial fishers. And so, before Rotondella takes clients out to a beach to fish for sharks, he makes it clear that if anybody does snag a shark, the fish will be freed after its catch. Moreover, Rotondella does not use barbed hooks. After a client catches a shark and reels it in to shallow water, Rotondella wades out to the shark and cuts the line holding the hook, enabling the shark to easily spit out the hook and swim off after suffering just a minor wound. He says, "The point is, for this [beach] to be a sustainable fishery you need to have fish swimming off and living a happy life."[39]

> **"The point is, for this [beach] to be a sustainable fishery you need to have fish swimming off and living a happy life."[39]**
>
> —A.J. Rotondella, sport fishing guide

Rotondella's experience illustrates how attitudes about sharks can be changed. After all, it took a decade of lobbying by environmentalists before the government of Costa Rica adopted a law protecting sharks. And although critics believe the US Shark Fin Sales Elimination Act does not go far enough to protect sharks, NOAA has nevertheless succeeded in prosecuting several violators. These actions mean sharks can be saved, but it will continue to take hard work by environmentalists as well as research by marine biologists who are tasked with protecting these valuable, and misunderstood, creatures of the sea.

# SOURCE NOTES

## Introduction: The Challenge of Saving Sharks

1. Quoted in Remy Tumin, "Tamayo Perry, Professional Surfer, Dies After Shark Attack Off Oahu," *New York Times*, June 24, 2024. www.nytimes.com.
2. Quoted in Jon Haworth and Matt Rivers, "Shark Attack Victim Tamayo Perry Remembered by Widow and Friend," ABC News, June 26, 2024. https://abcnews.go.com.
3. Quoted in *The Pulse*, "Sharks—from Fear to Fascination," National Public Radio, July 26, 2024. https://whyy.org.
4. Quoted in ABC News, "10-Year-Old Boy Describes How He Survived Shark Attack," March 30, 2023. https://abcnews.go.com.
5. Quoted in ABC News, "10-Year-Old Boy Describes How He Survived Shark Attack."

## Chapter One: The Unique Qualities of Sharks

6. Greg Skomal, *The Great White Shark Handbook*. Nashville, TN: Cider Mill, 2023, p. 20.
7. Melissa Cristina Márquez, "Shark Cannibalism: It's a Thing and It Just Got Weirder," *Forbes*, December 29, 2018. www.forbes.com.
8. Quoted in *The Pulse*, "Sharks—from Fear to Fascination."
9. Quoted in Deborah Byrd, "Sharks' 6th Sense Tuned to Attack," EarthSky, June 6, 2018. https://earthsky.org.
10. Quoted in Harry Baker, "Sharks Do Actually Sleep, and Sometimes with Their Eyes Wide Open," LiveScience, July 5, 2022. www.livescience.com.

## Chapter Two: Why Are Sharks Vital to the Environment?

11. Quoted in Titus Kakembo, "Lions Helping to Cope with Climate Change," New Vision, November 23, 2023. www.newvision.co.ug.

12. Quoted in *The Pulse*, "Sharks—from Fear to Fascination."
13. Lawrence Chlebeck, "Why Are Sharks Important to the Ecosystem?," Shark Champions, September 28, 2022. https://sharkchampions.org.au.
14. Florida International University Institute of Environment, "Ecological Importance of Sharks," 2024. https://environment.fiu.edu.
15. Quoted in *The Pulse*, "Sharks—from Fear to Fascination."
16. Alexander Nicolas, "Sharks Are Key to the Health of Our Oceans and Climate," World Wildlife Fund, 2024. www.worldwildlife.org.

## Chapter Three: What Are the Threats to Sharks?

17. Quoted in the Kennedy Center, "Steven Spielberg," 2006. www.kennedy-center.org.
18. Quoted in Jacquelyne Germain, "Steven Spielberg Regrets How *Jaws* Impacted Real-World Sharks," *Smithsonian*, December 20, 2022. www.smithsonianmag.com.
19. David McGuire, "How *Jaws* Influenced Shark Perception," Shark Stewards, 2023. https://sharkstewards.org.
20. Quoted in Ania Savage, "Jaws Spurs Shark Fishing," *New York Times*, October 12, 1975, p. 150.
21. Quoted in Savage, "Jaws Spurs Shark Fishing," p. 150.
22. Quoted in Marlowe Hood, "Many Sharks Closer to Extinction than Feared: Red List," Phys.org, March 22, 2019. https://phys.org.
23. Quoted in *The Pulse*, "Sharks—from Fear to Fascination."
24. Quoted in Krista Mahr, "Shark-Fin Soup and the Conservation Challenge," *Time*, August 9, 2010. https://time.com.
25. Mahr, "Shark-Fin Soup and the Conservation Challenge."
26. Karl Wasson, "The Pros and Cons of Shark Leather Boots," Stridewise, May 17, 2020. https://stridewise.com.
27. Lauren Evashenk, "The Truth Behind One of the Cosmetic Industry's Deadliest Ingredients: Squalane," Axiology, August 17, 2018. https://axiologybeauty.com.
28. Quoted in Jon Chapman, "'Never in My Wildest Dreams': A Night of Shark Fishing That No One Will Forget," *Bradenton (FL) Herald*, July 31, 2022. www.bradenton.com.
29. Quoted in Chapman, "'Never in My Wildest Dreams': A Night of Shark Fishing That No One Will Forget."

## Chapter Four: What Can Be Done to Save Sharks?

30. Quoted in Elizabeth Claire Alberts, "Still on the Menu: Shark Fin Trade in US Persists Despite Ban," Mongabay, December 27, 2023. https://news.Mongabay.com.
31. Quoted in US Department of Justice, "Owner of Japanese Fishing Vessel Pleads Guilty to Unlawful Trafficking of Shark Fins," October 9, 2020. www.justice.gov.
32. Quoted in Manuela Andreoni, "Many Laws to Protect Sharks Have Backfired, Researchers Find," *New York Times*, January 12, 2024, p. 8.
33. Quoted in Maxwell Radwin, "Costa Rica Announces Ban on Fishing Sharks," Mongabay, February 10, 2023. https://news.Mongabay.com.
34. Bloom Association, *Comparison Report, 2019–2020*. Hong Kong: Bloom Association, 2021, p. 5.
35. Quoted in SharkProject, "Austria Plans to Ban the Import of Commercial Shark-Products," December 7, 2022. ww.sharkproject.org.
36. Quoted in Shark Stewards, "Newsom Signs Chum Bill to Protect White Sharks," 2022. https://sharkstewards.org.
37. Quoted in *The Pulse*, "Sharks—from Fear to Fascination."
38. Quoted in *The Pulse*, "Sharks—from Fear to Fascination."
39. Quoted in *The Pulse*, "Sharks—from Fear to Fascination."

**Animal Welfare Institute (AWI)**
https://awionline.org
AWI heads many programs aimed at protecting animals it regards as endangered by human activities. Visitors to the organization's website can find AWI's list of restaurants in the United States and Canada serving shark fin soup by accessing the "Our Programs" link on the website. Volunteers who visit restaurants update the list if they find shark fin soup on the menus.

**Center for Rescued and Endangered Marine Species**
www.cremacr.org/en
The Costa Rica–based environmental group, which won a ban on hammerhead shark fishing along the nation's beaches, has many other campaigns underway to further protect species of sharks. Visitors to the group's website can find updates on the center's campaign to ban businesses from exporting any shark-related products, food or otherwise, from Costa Rica.

**Florida Museum of Natural History**
www.floridamuseum.ufl.edu
The Gainesville, Florida, museum, maintained by the University of Florida, chronicles and exhibits many aspects of nature and the environment. The museum's records of shark attacks on humans can be accessed through the website's "Research" tab.

**National Oceanic and Atmospheric Administration (NOAA)**
www.noaa.gov
This federal agency funds studies of the world's oceans. NOAA is tasked with assessing the viability of animal life

on the planet and maintains lists of those species threatened by human activity, such as overfishing. Visitors can find species of sharks on the NOAA threatened and endangered lists by accessing the “Fisheries” tab on the agency’s website.

**Oceana International**

https://oceana.org

This environmental group campaigns for laws and changes in human culture that could help protect the oceans, such as ending overfishing and protecting habitats, including coral reefs. By accessing the “Resources” link on the organization’s website, visitors can find information on marine life that is threatened, including several species of sharks.

**Shark Research Institute**

www.sharks.org

The Shark Research Institute funds scientific studies of sharks and lobbies for legislation to protect sharks. By accessing the “Learn” tab on the institute’s website, visitors can find descriptions, with illustrations, of each member of every species of shark known to exist in the world’s oceans.

# FOR FURTHER RESEARCH


## Books

Helen Czerski, *The Blue Machine: How the Ocean Works*. New York: Norton, 2024.

Jasmin Graham, *Sharks Don't Sink: Adventures of a Rogue Shark Scientist*. New York: Pantheon, 2024.

John Long, *The Secret History of Sharks: The Rise of the Ocean's Most Fearsome Predators*. New York: Ballantine, 2024.

Greg Skomal, *Chasing Shadows: My Life Tracking the Great White Shark*. New York: Morrow, 2023.

## Internet Sources

Elizabeth Claire Alberts, "Still on the Menu: Shark Fin Trade in US Persists Despite Ban," Mongabay, December 27, 2023. https://news.mongabay.com.

Jon Chapman, "'Never in My Wildest Dreams': A Night of Shark Fishing That No One Will Forget," *Bradenton (FL) Herald*, July 31, 2022. www.bradenton.com.

Florida International University Institute of Environment, "Ecological Importance of Sharks," 2024. https://environment.fiu.edu.

Jacquelyne Germain, "Steven Spielberg Regrets How *Jaws* Impacted Real-World Sharks," *Smithsonian*, December 20, 2022. www.smithsonianmag.com.

*The Pulse*, "Sharks—from Fear to Fascination," National Public Radio, July 26, 2024. https://whyy.org.

# INDEX

# PICTURE CREDITS

Cover: gustavoquiroga/Shutterstock

6: DPPI Media/Alamy Stock Photo
10: Wildestanimal/Alamy Stock Photo
13: Mark Conlin/Alamy Stock Photo
15: Max Topchii/Shutterstock
22: Andriy Nekrasov/Shutterstock
26: frantisekhojdysz/Shutterstock
29: Scenics & Science/Alamy Stock Photo
33: Shawshots/Alamy Stock Photo
37: Andreas Altenburger/Shutterstock
41: gary corbett/Alamy Stock Photo
45: KIKE CALVO/Alamy Stock Photo
47: Tomas Kotouc/Shutterstock
51: John Lander/Alamy Stock Photo

# ABOUT THE AUTHOR

Hal Marcovitz is a former newspaper reporter and columnist who has written more than two hundred books for young readers. He makes his home in Chalfont, Pennsylvania.